María Antonia Jiménez Dávila
Ismara Zamora León
Irene Luisa del Castillo Remón

Chemistry and basic biomedical sciences

María Antonia Jiménez Dávila
Ismara Zamora León
Irene Luisa del Castillo Remón

Chemistry and basic biomedical sciences

Technological alternative

ScienciaScripts

Cover image: www.ingimage.com

This book is a translation from the original published under ISBN 978-620-3-58577-3.

Publisher:
Sciencia Scripts
is a trademark of
Dodo Books Indian Ocean Ltd., member of the OmniScriptum S.R.L Publishing group
str. A.Russo 15, of. 61, Chisinau-2068, Republic of Moldova Europe
Printed at: see last page
ISBN: 978-620-3-66274-0

Chemistry and basic biomedical sciences. Technological alternative

Chemistry and basic biomedical sciences. Technological alternative

Maria Antonia Jimenez Davila* http://orcid.org/0000-0003-2996-9715
Ismara Zamora Leon2 https://orcid.org/0000-0001-9372-3467
Irene Luisa del Castillo Remon3 http://orcid.org/0000-0003-3361-1003

University of Medical Sciences of Granma. Cuba.

*Author for email correspondence: mdavila@infomed.sld.cu

1. Degree in Chemistry. Master in Higher Education. Assistant Professor. Faculty of Medical Sciences of Manzanillo. University of Medical Sciences of Granma. Cuba. E-mail: mdavila@infomed.sld.cu
2. Degree in Spanish Literature. Master in Didactics of Spanish and Literature. Assistant Professor. Faculty of Medical Sciences of Manzanillo. University of Medical Sciences of Granma. Cuba. E-mail: ismazleo@infomed.sld.cu
3. Degree in Biology. Master in Higher Education. Assistant Professor. Faculty of Medical Sciences of Manzanillo. University of Medical Sciences of Granma. Cuba. E-mail: irecastillo@infomed.sld.cu

Table of Contents

SUMMARY 3
INTRODUCTION 4
CHAPTER I 8
METHODS 22
CHAPTER II 25
RESULTS 44
DISCUSSION 48
CONCLUSIONS 50
BIBLIOGRAPHICAL REFERENCES 51

SUMMARY

Rationale: the process of teaching and learning of chemistry requires constant updating in correspondence with the requirements of the training of health professionals.

Objective: to elaborate a technological alternative that allows to raise the level of learning of the chemical contents and its practical applications in the field of health.

Methods: an educational software was developed to support teaching for the linking of Chemistry with Basic Biomedical Sciences, for this Macromedia Flash was used as a tool for design and assembly of the different modules that make up the product. The technological alternative was implemented in the careers of Health Technology that received Chemistry in the course 2017-2018. A quasi-experiment was applied with an exit test for the experimental group. Theoretical methods were used: historical-logical, inductive-deductive, analytical-synthetic, systematic-structural-functional and empirical: documentary analysis, surveys, classroom observation, expert criteria. The participants signed the consent to collaborate and accepted the publication of the results with the guarantee of their anonymity.

Results: the results of the quasi-experiment show a qualitative leap in the learning of the students of the experimental group, in which it was achieved that most of them were located in a high level of learning to be able to establish the relation molecular cause-disease-practical application of the chemical contents.

Conclusions: the implementation in the educational practice of the technological alternative allowed to raise the level of learning of the chemical contents in its practical applications in the field of health.

Keywords: digital resource, chemistry, biomedical sciences, teaching and learning, teaching-learning

ABSTRACT

Foundation: the Chemistry teaching-learning process requires constant updating in accordance with the demands of health professionals' training.

Objective: to develop a technological alternative this allows raising the learning level of chemical contents and their practical applications in the health field.

Methods: Teaching support educational software was developed to link Chemistry with Basic Biomedical Sciences. Macromedia Flash was used as a design and assembly tool for different modules included in this product. The technological alternative was implemented in the Health Technology training programs that received Chemistry in the 2017-2018 academic years. A quasi experiment with an exit test was applied for the experimental group. Theoretical methods were used: historical-logical, inductive-deductive, analytical- synthetic, systemic-structural-functional and empirical: documentary analysis, surveys, class observation, and expert criteria. The participants signed the consent to collaborate and accepted the publication of its results with the guarantee of their anonymity.

Results: the results of the quasi-experiment show a qualitative improvement in the students learning in the experimental group. Most of them were placed at a high level of learning by establishing the relationship between molecular cause-disease-practical application of the chemical contents.

Conclusions: the implementation of this technological alternative in educational practice allowed raising the learning chemical content level in its practical application in the health field.

Keywords: digital resource, chemistry, biomedical sciences, teaching learning

INTRODUCTION

The current scientific and technical advances in the chemical sciences and their applications in medicine, agriculture and meteorology, among other branches, pose to society and to the contemporary school the need to attend in a different way to the learning and intellectual development of the new generations. (1)

The challenge of these times lies in training professionals capable, not only of processing a high volume of updated information, but who understand and act, with knowledge, personal involvement and responsibility for the solution of the problems that arise in everyday life, in such a way that they gradually appropriate the culture of humanity and act in accordance with the values that society demands. (2)

The teaching of chemical contents, both in the undergraduate and postgraduate formation, is facilitated with the use of teaching aids, which from a philosophical perspective are based on Marxist gnoseology, as they are imbricated in the process of knowledge construction, from the Leninist formula of knowledge, as visualizers of reality intervene in the cognitive process by providing the sensations and perceptions of objects and phenomena of that reality.

According to Garcia-Valcarcel's criteria, teaching means are understood as "the content carrier that materializes the actions of the teacher and the student for the achievement of the objectives". (3)

Nowadays, the use of computer resources as an educational technology for learning and teaching has gained remarkable relevance, since the learning of subjects and the development of skills can be facilitated through this way. Computer resources are also considered teaching aids. (4)

The use of computer in education presents positive characteristics such as interactivity, personalization, ease of use, research medium in the classroom, motivating medium, individual learning, so it should be used more to improve different learning. (5)

The search for resources to support the teaching and learning (T/L) of science, particularly chemistry, has been a constant task whose results have put at the service of the educational community a great number of elements: from heavy blackboards to practical electronic devices capable of carrying out a great number of tasks. Until 1929, the radio and projectors were the most popular tools in this context. At the same time (1930-1939) slides appeared and two years later a paper on the use of film in chemistry teaching was published. In 1956 television was used for the first time to broadcast closed-circuit chemistry classes (6).

In the 1970s to 1990s, microcomputers and personal computers were introduced, which ushered in the digital age and the Internet (1990 - present), with the development of software and digital resources that offer various options to motivate students to learn chemistry.

Nowadays there is a rapid development of technological tools and individuals who do not adapt to their pace of evolution, for political, social or economic reasons, may feel intellectually discriminated against. The application of Information and Communication Technologies (ICT) to the process of E/A arises as a necessity to help the full incorporation of young people to the Information and Knowledge Society (7).

The use of ICT in the classroom allows students to complement other forms of learning used in the classroom, improve understanding of concepts that are difficult or impossible to observe with the naked eye or in school laboratories, use representations to develop school projects with classmates and teachers. (8)

The educational software, in the classroom, offers the possibility of dynamically combining the use of different teaching media. Through the digital support, images and sounds of higher quality can be achieved and the student has greater possibilities to interact with the content. (7, 8)

Chemistry studies the various substances that exist on the planet, as well as the reactions that transform them into other substances. On the other hand, it studies the structure of substances at the molecular level; and, last but not least, their properties.

The research carried out on the Didactics of Chemistry and Biochemistry and the results divulged in national and international events in the context of the medical university are scarce. As a regularity of these investigations, which constitute limitations in the theoretical order, within the limits of this study, it is revealed that the theoretical contributions are directed to organize the teaching-learning process of Chemistry and Biochemistry with exercises, problems, the relations of compounds with life, however, the direction of learning with implications in the professional practice is not approached from the methodological point of view. [(9)]

An approach to the process of teaching-learning of chemistry allowed us to verify insufficiencies in the direction of the contents, which implies that the results that are obtained do not satisfy the purposes of the Ministry of Health, and that is evident in the insufficient quality of the services in the health care units. The difficulties detected can be summarized as follows: the methods and procedures used by the teachers still do not satisfy the linkage of the contents according to the relation molecular cause-disease-practical application in the field of health, the specialized bibliography of the subject for health careers is scarce, limited use of the educational technologies for the direction of the learning of the chemical and biochemical contents by the teacher and for the self-learning of the students in the careers of Health Technology, the students in the process of construction and application of the chemical contents still show difficulties to emit judgments, criteria and points of view on the health problems that the community faces, their causes and possible solutions.

The analysis of these insufficiencies allowed us to identify that there is insufficient bibliography that addresses the link between chemistry and basic biomedical sciences to facilitate the learning of chemical contents in their practical applications in the field of health.

For these reasons this research was carried out with the aim of developing a technological alternative to raise the level of learning of chemical content and its practical applications in the field of health.

CHAPTER I

CHARACTERIZATION OF THE TEACHING-LEARNING PROCESS OF CHEMICAL CONTENTS IN THE CHEMISTRY PROGRAM FOR HEALTH TECHNOLOGY CAREERS.

Theoretical references of the teaching-learning process of the chemical contents in the subject Chemistry in Health Technology careers. From a philosophical perspective, the Marxist conception of the world, in particular, the principle of the universal concatenation of the phenomena and the dialectic relation between the philosophical categories cause-effect, allow to base the necessity of establishing relations between the chemical contents and their practical applications in the field of health, which finds its expression in the relation structure-property-practical application. The objectivity in the treatment of these relations constitutes a demand and a necessity in the direction of the teaching-learning process, as it takes into account the properties and characteristics of the objective reality itself, which is highly changing and complex [(10)].

Likewise, chemistry contributes to the formation of a scientific conception of the world by making it possible to study and understand the molecular foundations and biochemical processes that occur in organisms, the structure-property-function relationship of the chemical compounds characteristic of living beings, as well as the chemical transformations that occur in them and, in addition, the molecular mechanisms involved in the regulation of such transformations, all of which has Marxist Gnoseology[(11)] as its methodological basis.

A didactic analysis of the teaching-learning process of chemistry reveals that in it there are contradictions inherent to any educational process, among which is considered essential the contradiction between the theoretical and practical problems that students are required to solve and the real possibilities they have to solve them [12, 13].

An analysis of the didactic categories of the teaching-learning process of chemistry reveals that its objective, in the training of health technologists, is directed to the establishment of the structure-property-function relationship of chemical substances in correspondence with the professional profile, which determines as essential content, the concepts related to chemical compounds, their structure, classification, nomenclature, isomerism, physical and chemical properties and functions; The latter allows its linkage with the professional practice of technologists in the field of health. [14]

The teacher should select the part of the content in which the student can express his creative possibilities, to this contributes the system of didactic procedures and means for the education of an active personality.

To successfully meet these requirements it is important to use active methods that promote meaningfulness in student learning, to the extent that they allow to establish the relationship of the content with its practical applications. [15]

In the context of this research, the assumption of referents in the didactic order for the construction of the proposed alternative cannot be separated from the analysis that the teaching media deserve, since they occupy an essential place in the relationships that are established.

At present, a wide range of definitions of teaching aids can be found in the pedagogical literature [16]. This is nuanced by the positions of different specialists on the role of media in the theory of teaching (also called general didactics) and in the direction of the pedagogical process, which can be summarized as follows:

- Some do not consider them to be included among the fundamental didactic categories of the theory of teaching as a scientific branch of pedagogy.
- There are different criteria for their study: some authors refer to the means

of teaching when dealing with the content of the didactic principle of the unity between the concrete and the abstract (also called the visualization of teaching), others include them as a factor of direction of the teaching methods in the classroom action, while for other specialists they acquire the status of an integral component of the teaching process.

- Another important aspect, derived from the predominant conception of learning in the planning and execution of the teaching process, is related to the very denomination of this component: the means. Under the recognition of the humanistic and personological character of the process, objectives and methods of learning and teaching, it would be consistent to refer to learning media and teaching media insofar as they are incorporated as material support to the performance of teaching tasks of one or the other, that is, of students and teachers. Given that the same medium can appear with different functions depending on the subject on whom it is used, the denomination teaching medium will be adopted only with the purpose of homologating the terminology with the one that appears in the pedagogical literature in a generalized way. This criterion, purely conventional, does not contradict at all the points of view and conceptions assumed with respect to the functions of the students in the teaching process and learning in particular.

Pedagogues and didacticians define teaching aids in the existing theories in many ways; some by their pedagogical functions, others by their physical nature and many without stating classificatory criteria.

According to the criteria of Dr. Fernandez Rodriguez, quoted by Lombillo Rivero, teaching means are understood as "the bearer of content that materializes the actions of the teacher and the student for the achievement of the objectives". (17)

From the point of view of communication theory, the teaching media are the channel through which teaching messages are transmitted. According to

Gonzalez Castro, quoted by Peraza Zamora and others, the teaching media are the components of the teaching-learning process that act as the material support of the methods to achieve the stated objectives [(18)].

These scholars of the subject formulate definitions of teaching media, however, the endogenous development achieved by science and technology in the conditions of today's world have led to the emergence of other media, including audiovisual and magnetic media that are not relevant in these definitions, so in the context of this research, In correspondence with the demands of the proposed objective, it is necessary to consider in addition, those elements of virtual nature, such as educational software, carriers of cognitive messages that illustrate the structure and essence of the movement of objects, phenomena and processes of reality that must be assimilated by students to achieve the proposed objectives.

In such a way, the teaching media have been considered not only as the static representation of knowledge, nor the natural movement of objects, but also the movement that can be simulated from a virtual representation, as in the case of videos and educational software with the use of computers. (19)

From a philosophical perspective, the teaching media are based on Marxist gnoseology, since they are imbricated in the process of knowledge construction, from the Leninist formula of knowledge, as they intervene in the cognitive process by providing the sensations and perceptions of the objects and phenomena of that reality as visualizers of reality.

Organizing the work with teaching aids requires considering three essential phases: selection, design and utilization. In the first one it is necessary to consider from the analysis of the content and the method, which are the suitable media for the direction of learning; in the second one, teachers and students intervene depending on the signs and symbols that will be used to clarify the cognitive messages as forms, color, size, lettering, relation figure-background, margins among others. The utilization includes the aspects of the handling of the medium by teachers and students such as: moment of use, place of placement, time of permanence or exposition, elements to be used to concentrate the attention of the student, relation word-image. [20]

The previous approaches do not exhaust the conceptualizing arguments of the media, but they are the referents to elaborate an explanatory framework of the relations between the media and the subject who learns, starting from these and in the context of the strategies and alternatives of teaching and learning. The means of teaching are conditioned by the objectives, the contents and their peculiarities, the methods used and the organizational forms of the teaching activity and this, in turn, conditions the instruments of measurement of learning that are used; therefore, the place of the means of teaching in the process conditions them as components of the system in relation to the other didactic components.

From the psychology of teaching and learning, it can be considered that the materialistic tradition of Russian Naturalistic Psychology and Pavlovian theory together with the rise of the Soviet socialist state laid the foundations for the emergence of Materialistic-Dialectical Psychology.

In the context described above, Vigotsky is considered to be the initiator of the so-called cultural-historical approach that, with high productivity and theoretical excellence, constitutes a fairly complete elaboration of the education-development relationship from Marxist-oriented positions. [21]

A relevant idea in this approach is related to the concept of Mediation, which conceives the relationship between the subject (learner) and the object (culture) as a dialectical interaction (subject - object) in which there is a mutual transformation mediated by socio-cultural instruments in a given historical context. For this author, there are two forms of mediation: the influence of the socio-historical context (adults, peers, organized activities,) and the socio-cultural instruments used by the learner (tools and signs) which in the context of this research the designed software constitutes a mediation instrument.

The qualitative transformations that show psychical development are related to changes in the use of instruments as a form of mediation, which enables the subject to perform more complex, qualitatively superior actions on objects.

Appropriation, according to Vigotsky, constitutes the fundamental mechanism by which human psychic development takes place and, in fact, becomes a new psychological category, which surpasses paradigms that gave a less dialectically complete interpretation of this process.

The socio-cultural instruments, with their cultural nature and with the expression of the human social essence, give a new dimension to the solution of this problem of a philosophical nature for psychological science. In this sense, these ideas were influenced by dialectical materialism and this is very clearly manifested in their theoretical and methodological conceptions.

According to this author, the problem of knowledge between the subject and the object of knowledge is solved with the dialectical interactionist approach (subject - object) in which there is a relationship of reciprocal influence between the two; this interaction in double direction is called object activity as it transforms the object (reality) and the bearer of the activity itself, the subject (man). In the object activity, the historico-social practices (the production process) are materialized and developed.

In this sense, from the Marxist interpretation, there is a dialectical leap with respect to the theories that understand the activity of the subject as a pure individual and biological adaptation, towards a conception where the activity is seen as a social practice subject to the historico-cultural conditions. Therefore, the relationship between the subject and the object of knowledge is mediated by the activity that the subject performs on the object with the use of socio-cultural instruments, which can be basically of two types: tools and signs. Each of these instruments orients the subject's activity in a different way. The use of tools produces transformations in the objects or as Vigotsky would say, the tools are externally oriented. On the other hand, the signs produce changes in the subject who carries out the activity, that is to say, they are internally oriented.

The computational media constitute sociocultural tools, the activity that mediates is learning, while the subject is the student and the object is the morphophysiological processes, the signs correspond to the labels that the students must unravel in interaction with communication. [(22)]

Through mediated activity, in interaction with its sociocultural context, the subject constructs, internalizes higher psychological functions and consciousness. This learning comes into contact with meaningful learning to the extent that it enhances the establishment of relationships between new content and the affective and motivational world of students, relationships between the concepts already acquired and the new ones that are formed, relationships between knowledge and life, between theory and practice. From this meaningful relationship, the content of new learning acquires a real value for the learner, and increases the chances that such learning is durable, recoverable, generalizable, transferable to new situations.

Vigotsky, emphasized the social nature of the process of internalization given as a psychological mechanism of appropriation, by pointing out the decisive role of the adult as mediator of the subject-object relationship and bearer of the most general and concrete forms of the historico-social experience and culture contained in the objects of the reality surrounding the subject.

From this conception arose his notion of "zone of proximal development" of great value for the understanding of psychic development, Differential Psychology and Pedagogy. This zone is determined by what the subject can accomplish in collaboration, under the direction and with the help of the adult or even his own peers. It offers a measure of the potentialities of psychic development. [2]

Vigotsky's approach had extraordinary consequences for pedagogy in that it made it possible to rethink the problem of the relation between teaching and psychic development; to establish how teaching and education guide and lead development, to oppose his approach to the most generalized of his time, in which teaching depends on maturation (Piagetian conception), or to the parallelism between teaching and development (behaviorism).

From the technological point of view, nowadays, the use of computer resources as an educational technology for learning and teaching has acquired remarkable relevance, since the learning of subjects and the development of skills can be facilitated through this way. These are also considered teaching aids.

The use of the computer as a teaching medium is materialized through the use of a wide range of types of programs, which can be used in the process of acquisition or active consolidation of knowledge, by the student, with varied approaches. [24] Each of them has specific purposes, aimed at contributing to the development of different functions of the teaching process.

It is important to be sure that the medium to be used is the most effective in the process of assimilation of knowledge and not the one that is used from a simplistic position and little analytical from the didactic, in addition, should not miss the pedagogical orientation, proper treatment of the contents to be treated according to the curriculum and subject programs. Hence the importance of a correct selection by the teacher of the means to be used in their classes.

It should be taken into consideration that the software is conceived to be used within a teaching activity in two ways: in the classroom and in the extra-class activity of the students. In both cases it is fundamental the orientation and direction of the teacher for its use.

The educational software, in the classroom, offers the possibility of dynamically combining the use of various teaching media. Through the digital support can be achieved images and sounds of higher quality and the student has greater opportunities to interact with the content.

In order to use software as a teaching medium successfully, the following elements must be taken into account:

- ❖ Make a careful selection or elaboration in correspondence with the educational needs, its quality and the teacher's management of the learning environment.

- ❖ To ensure the material conditions in terms of availability of equipment, computational means compatible with the installed equipment and the quality required to solve the difficulties.

To ensure the conditions in relation to the indispensable minimum formation of the computer skills of teachers and students. (25)

The use of teaching aids of this nature is justified by the advantages they offer, which can be summarized as follows (26)

Interest. Motivation. As a reflection of the student's actuation, in which they express psychic configurations of their personality as the motivational orientation that moves them in search of a conscious objective, and the state of satisfaction that sustains them in their actuation, as it keeps them motivated towards the activity, when using the resources of the ICT as motors of the learning, since it incites to the activity and to the thought.

Interaction. Continuous intellectual activity. Students are permanently active when interacting with the computer and with each other at a distance. They maintain a high degree of involvement in the work. The versatility and interactivity of the computer, the possibility of dialoguing with it, the large volume of information available keep their attention.

Cooperative learning. The tools provided by ICT (information sources, interactive materials, e-mail, shared disk space, forums) facilitate group work and the cultivation of social attitudes, exchange of ideas, cooperation and personality development. Group work stimulates its components and makes them discuss about the best solution for a problem, criticize, communicate their discoveries. Moreover, fatigue appears later, and some students reason better when they see someone else solve a problem than when they have the responsibility themselves.

High degree of interdisciplinarity. Educational tasks carried out with computers allow for a high degree of interdisciplinarity, since the computer, due to its versatility and large storage capacity, allows for very different types of treatment to a very wide and varied information.

Digital and audiovisual literacy. These materials provide students with a contact with ICT as a means of learning and a tool for the process of information (access to information, data processing, expression and communication), generator of experiences and learning. They contribute to facilitate the necessary information and audiovisual literacy.

Development of information search and selection skills. The large volume of information available requires the implementation of techniques that help the localization of the information needed and its valuation.

Improving the skills of expression and creativity. The tools provided by ICT (word processors, graphic editors) facilitate the development of written, graphic and audiovisual expression skills.

Easy access to a lot of information of all kinds. They make available to students and teachers a large volume of information (textual and audiovisual) that can facilitate learning.

Within the classifications of educational software that can be used in the teaching-learning process of a particular subject is the multimedia that, according to Rodriguez Lamas, R "... has the possibility of accessing the different contents that are offered in the software with the use of different media, and, in this way, develop their knowledge in a more interactive way". (27, 28)

In the case of this research we assume the definition exposed in the previous paragraph and that founds the conception of the computerized teaching medium designed.

The causes that originate the need to use this medium can be summarized as follows:

- Insufficient learning of chemical contents.
- Formation of an active professional activity.
- Existence of a large amount of scientific information, derived from the scientific-technical revolution, which generates contradictions between the contents to be taught and the methods, procedures, and means of teaching.
- Use of traditionalist methods that do not promote the assimilation of knowledge.

Learning situations are shown in the middle. This approach, due to its heuristic character, makes it possible for the direction of the teaching-learning process to mobilize the students' thinking if it begins with the presentation of learning situations that generate problems or questions, for which conditions must be created that foster a motivational environment that promotes the students' interest in learning the chemical content.

The aim of these situations is to generate interest in the search for relationships between the characteristics of the disease described and its explanation from the molecular point of view, which facilitates the learning of chemical contents by taking into account the relationship between molecular cause-disease-practical application in the field of health.

The aforementioned implies finding a path that is not known beforehand, that is, the search for an alternative with which to find a solution; this requires previous knowledge and skills, through which new knowledge is constructed. Therefore, problematization creates learning environments that allow the formation of autonomous, critical subjects; in addition, it allows them to acquire ways of thinking, habits of perseverance, curiosity and confidence in unfamiliar situations that serve them in other contexts of action, including the workplace.

The questions related to the learning situation highlight the most significant or interesting unknowns that can be derived from the situation and that require an explanation by the students; these questions break down the learning situations posed into their essential aspects and facilitate their solution by the students.

It is possible to demonstrate, in the first place, that the chemical contents can be linked to aspects of the practical life of Health Technology professionals, as long as their teaching is conceived through a problematizing approach, for which the teaching-learning process must be organized with a system character and the tools that allow its implementation must be elaborated.

The fundamental formative objective of the teaching of chemistry is to prepare students with knowledge about the composition, properties and structure of substances which leads to the questioning about the treatment or not of the practical applications of these compounds in their link with the structure-property-function relationship, so that their learning is significant from its impact on the future professional life of graduates of the careers of Health Technology.

The criteria contributed by Rojas Arce, in his book "Metodologia de la ensenanza de la Quimica" ([29]), are assumed in this research from the didactic point of view as referents, when considering that the practical applications of the compounds must be adjusted to the health environment. These judgments can be summarized as follows:

- Study of the structure of compounds.
- Treatment of the properties.
- Comparison of compounds and classification.
- Practical applications of composites.

In accordance with the above, it is required the application in the teaching-learning of Chemistry for the formation of health technologists, of technological alternatives (which includes questions to reveal knowledge, unsolved tasks with different ways of solution, assumption and defense of positions, approach of learning situations) that require the student to reflect, to search independently for knowledge, to establish conclusions, and that allow him in the same measure that he assimilates generalizing procedures of mental work, by the own conception of the task, to observe, compare, generalize, elaborate, develop and generalize, the independent search of knowledge, the establishment of conclusions, and that allow the student to observe, compare, generalize, elaborate concepts, make assumptions, solve and conceive new problems, argue, evaluate, as well as assimilate generalizing procedures of mental work, by the conception of the task itself. From a developmental perspective, it is necessary to focus the teaching-learning process of chemistry as an interactive process of assimilation, from more participatory, conscious, empowering, motivating, reflective and generalizing dimensions, in a word, from a meaningful dimension.

Any developmental teaching-learning process requires promoting the protagonism of the students, for which the teaching of theoretical-practical-experimental character as in the case of the teaching of chemistry, In this sense, Silvestre Orasma and Zilberstein-Toruncha [(30)] propose a system of didactic requirements, which in the context of this research are feasible to use in the teaching-learning process of chemistry.

Being consistent with these ideas, learning should be stimulated from the diagnosis of the current development achieved by each student at each stage, in order to promote the proximate development, whose level will be measured by the actions that the student can perform by himself, who previously will be supported by the teacher, the group, family and community; that is, that independent work is encouraged in the student so that he can operate with the knowledge by himself, in new situations.

If the chemical knowledge to be taught were independent of each other and if they all had the same level of complexity, it would make little difference what organization or order of treatment they should be given.

If the way in which they are best acquired were to follow the pattern established by the logic of the discipline, or the strict order in which they have been constructed by scientists throughout the history of thought, there would be little interest in analyzing the sequence of contents and, in any case, it would not be a problem.

Until now, the organization of chemistry courses has been traditionally conceived by a rigid structure, which is framed in the programs and orientations that govern the teaching of chemical sciences, and on the other hand, teachers who do not seek solution to the growing need to achieve an adequate understanding of the content, do not comply with principles of this science and do not contribute to foster in students an adequate scientific conception of the world.

From another perspective, the meaningful learning proposed by the educational psychologist David Ausubel in the 70's, conceives the student as an active processor of information by stating that: "Meaningful Learning comprises the acquisition of new meanings and, conversely, this is a product of Meaningful Learning"; he also states that: "... the educational process (...) is the human mechanism par excellence for acquiring and storing the amount of ideas and information represented by any field of knowledge", and he attributes it to the intentionality and substantiality of the learning task with the cognitive structure.) is the human mechanism par excellence to acquire and store the amount of ideas and information represented by any field of knowledge", and attributes it to the intentionality and substantiality of the learning task with the cognitive structure, that is, that learning is meaningful when there is a relationship between the new knowledge or learning material and the student's previous knowledge [(31)]; these postulates are well applicable to the teaching-learning process of chemistry in the context of the Cuban university as long as one considers, in addition, the relationship between the instructive and the affective, in the learning of biomolecules, based on the Vygotskyan postulates.In the teaching-learning process of chemistry it is required that these aspects are taken up again for the conception of alternatives for the learning of chemical contents, whose methodological systematization would help the students to face the process with solid and lasting knowledge, besides that the teacher can use them to establish links with the new knowledge and offer the ways to relate them with the demands of their formation as future professionals.

The use of computational means, allows the development of thinking, as it requires analysis, abstraction, reflection and generalization (general intellectual skills); it contributes to the development of autonomy and critical reasoning while enabling the technologist to understand the practical applications of these contents. Considering the complex character of the training process of the professional of the careers of Health Technology, in its dynamics, the instructive and the problematizing must be integrated so that he can act and proceed according to the complexity of the educational reality and the demands of his functions and the professional profile.

From the theoretical analysis carried out in this epigraph it is inferred that the learning of chemical contents in the training of health technologists through the process of teaching-learning of chemistry requires a process that is developmental and interactive, in which the learning situations that are used, promote significance and meaning for the student, systematized by means of methods, procedures and means that stimulate the development.

METHODS

In the investigation, the study units are coincident, they were represented by students of the different careers of Health Technology that receive Chemistry in their study plans at the University of Medical Sciences of Granma in the academic year 2017-2018 among which were: Clinical Bioanalysis, Hygiene and Epidemiology, and Imaging and Medical Radiophysics, among all 45 students, The type of study is experimental and was developed as a quasi-experiment modality.

The independent variable was constituted by the technological alternative (software) and the dependent variable in this case is the level of learning of the chemical contents in its practical applications in the health field. The operationalization of the dependent variable was carried out from its conceptualization and the establishment of dimensions and indicators that made possible the measurement of the transformations operated in it by the action of the independent variable. The following dimensions were established: cite examples of chemical substances used in the profession, explain the structure-properties-functions relationship of the substances and exemplify the application of the chemical content, taking into account the structure-properties-functions relationship in their professional practice. For each dimension, indicators were established that placed the students in a high, medium and low level of learning of the chemical contents in their practical applications in the field of health.

The research process was achieved through the use of different methods[10].From a dialectical-materialist position, research methods were applied, both theoretical and empirical, which are related to continuation: historical-logical: the regularities and tendencies of the teaching-learning process of chemistry in the careers of health technology were established; analytical-synthetic and inductive-deductive: its use allowed to establish the order of the content to be included in the product and in the processing and interpretation of the data to establish conclusions; systematic-structural-functional: the structure of relationships between the contents that make up the proposed alternative was formed; modeling: Macromedia Flash was used as a tool for design and assembly of the different modules that make up the product; observation (in classes): the characteristics of the teaching-learning process of the chemical contents were verified, taking into account their practical applications in the field of health; surveys (to professors who teach chemistry): allowed to know the particularities of the content of the classes and to determine the contents with greater difficulty; documentary analysis (of study plans, teaching programs, attendance records and evaluations): allowed to characterize the fundamentals of the teaching-learning process of chemistry from the educational policy in the search of documentary evidence that corroborate the declared inadequacies; evaluation by expert criteria: the feasibility of the proposed technological alternative and its possible impact on the learning of chemical contents was evaluated; pedagogical experiment: allowed to assess the effectiveness of the technological alternative in the direction of the teaching-learning process of chemical contents. Techniques of descriptive and inferential statistics were used for the treatment, interpretation and evaluation of the results of the application of the empirical methods.

The results of the applied empirical methods demonstrate the need to establish a way to link chemistry with basic biomedical sciences in such a way that the student can appreciate and apply this knowledge throughout his career and in his professional performance. In the process of construction of the technological alternative, specialists were consulted in order to improve this proposal. Also, in a second moment, before its application, a consultation was carried out to determine the feasibility of the application of the alternative in the careers of Health Technology, the results of which are exposed below:

Twenty experts were selected, by means of the methodology of preference, from a total of 25 candidates, these were chosen from the teaching staff that teaches the contents of the subject Chemistry, as well as from others related to the professional profile of these careers; in addition, specialists from the University of Medical Sciences, the University of Pedagogical Sciences and the University of Granma were consulted. The selected experts hold the main teaching categories of Assistant Professor, Assistant Professor and Professor, in addition, among them, there are Masters in Sciences and Doctors in Pedagogical Sciences.

This step allowed, by means of parametric and non-parametric techniques, the calculation of Kendall's coefficient of concordance (W). For this purpose, the statistical package SPSS (StatisticalPackagefor Social Sciences), version 15.0, was used.

The non-parametric test was applied for K samples related to the judgments made by the experts.

The study complied with the II Declaration of Helsinki and Cuban ethical regulations. The data obtained, the grades of the final exams, the surveys applied and the criteria expressed by the participants were protected under the precept of confidentiality. The participants signed the consent to collaborate with the research and accepted that the results achieved would be published with the guarantee of preserving their anonymity.

CHAPTER II

DESCRIPTION OF THE PROPOSAL

The interaction of the student with the computational means makes possible to integrate in a dialectic way in the teaching-learning process, the interactive elements of the educational software with an integrating approach of the cognitive and the affective in the learning of the regulatory systems in the human body, in this way, the use of software as sociocultural instruments is oriented to produce changes in the learning of the students.

For the proposal of the technological alternative it was assumed the utilization of the educational software that can be used in the teaching-learning process of the subjects as a means of teaching, because it offers the possibility of accessing the different contents that are offered with the utilization of different media, and in this way develop their knowledge in an interactive way, for which it is fundamental the orientation and direction of the teacher for its utilization; according to the postulates of Rodriguez Lamas [32] in his work "Introduction to the Educational Informatics". In addition, it is assumed as advantages:

- ❖ Interactivity, dynamism and motivation.
- ❖ Facilitates animated representations.
- ❖ It affects the development of skills through systematization.
- ❖ It allows to simulate complex processes.

❖ It reduces the time available for teaching a large amount of knowledge by facilitating differentiated work, introducing the student to work with computerized media.

❖ It facilitates independent work and the construction of new knowledge.

❖ It strengthens collaborative learning, the ability to formulate problems, the tendency towards self-training.

Among the aspects mentioned above, interactivity is a characteristic of the New Technologies and therefore deserves a timely comment, is the possibility offered by this technology so that, in the direct user - machine relationship, can be exchanged at a given time a greater volume of knowledge and knowledge of science that are difficult to show the student. With the use of this technology in the teaching-learning process of the subject, the students feel a greater motivation since, with the presentation of multimedia information, it is more enriching and enjoyable, which increases the use and attractiveness of the system for the students who use it. In its interactivity with the computational teaching medium, the visualization of the structure and composition of the molecules with detailed images and videos is made possible.

The design of the graphic elements of the interface was carried out with the use of Adobe Photoshop and the programming was supported by Macromedia Flash and Macromedia Director MX 2004. The use of these authoring tools has allowed us to achieve a high level of design combined with complex animations that contribute to the achievement of the objectives of utilization of the computational medium.

DESCRIPTION OF THE TECHNOLOGICAL ALTERNATIVE

The software contains the contents of the program of Chemistry for the careers of Health Technology and a second part in which articles are collected where the connection of Chemistry with the basic biomedical sciences is evidenced, element that is included to motivate and to demonstrate the social necessity of this learning and therefore for its profession [(33)].

We made a software (technological alternative) with the following screens: the main screen has three parts: in the upper part the title of the product is shown and in the right part, three action buttons, Help, Minimize and Close; in the central left part there are two buttons: one of them is Articles and when you click on it you can access to the main topics of the product; the other button is Galenas, and through it you can access to the Galenas of Images and Videos; in the lower part there is the Sound button where you can activate or deactivate the background sound. This design is similar in all the screens.

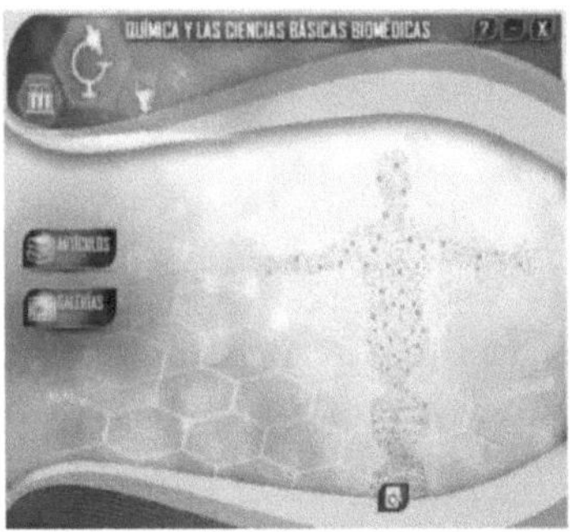

Main Screen

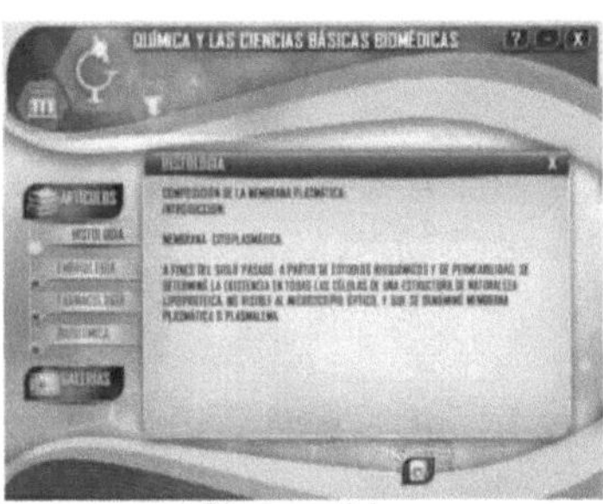

Display articles

Displays the Item and Galenas buttons, clicking on the Item button options displays the product topics. The selected theme is displayed in the middle content box and the name of the theme is displayed on the top of the theme. The selected option is set to a more vivid color, the unselected ones are set to transparent. The articles shown demonstrate the relationship of chemistry to the basic biomedical sciences. They are classified by discipline.

Embryology:
Biological membranes

Histology:

Composition of the plasma membrane:

Pharmacology:

Components of the cell and their transport mechanisms with the pharmacokinetics of drugs.

Pharmacological Receptor Theones

Biochemistry

Molecular diseases and their relationship with cellular molecular biology.

Lipids and their importance in the human organism.

Physiology

Composition of saliva.

These articles show the importance of chemistry in the different biomedical specialties.

For example for the specialty of Pharmacology the following article is shown:

Components of the cell and their transport mechanisms with the pharmacokinetics of drugs.

Introduction

Pharmacology is the unified study of the properties of chemical agents (drugs), living organisms and all aspects of their interactions.

Cellular and molecular biology is precisely the plan of unified organization; in other words, it is the analysis of the molecules and cells that constitute the building blocks with which all living forms are constituted.

Biochemical studies have shown that living matter is composed of 105 the same elements that constitute the inorganic world, although there may be fundamental differences in their organization. In the inanimate world there is a continuous tendency to a thermodynamic equilibrium with a random contribution of matter and energy, while in a living organism a high degree of structure and function is maintained during the transformation of energy and matter, based on the constant input and output of these. Biochemists isolated from complex mixtures of cellular constituents, in addition to inorganic components, much more complex molecules, such as proteins, fats, carbohydrates and nucleic acids. These biochemical studies also demonstrated the existence of a basic unity in the whole world.

The mode of action of pharmaceuticals is not generally understood by people outside certain health-related professions. However, public awareness and concern about the intermittent or daily exposure of large sections of the population to pharmaceuticals and other chemical compounds is greater today than ever before. At the beginning of the 19th century a series of important revolutionary changes in the field of chemistry occurred almost simultaneously with those in physiology that allowed the elucidation of the dynamic actions of chemical compounds on biological processes. They provided the means to discover what exactly drugs do in the living organism and established the principle that the chemical structure of drugs determines their effect on the organism.

The aim of the work is to demonstrate the link between pharmacology and chemistry through the different components of the cell and its transport mechanisms with the pharmacokinetics of drugs.

Development

Structure of a cell:

Cells, although very small, are very complex, and are made up of various structures that allow them to carry out functions such as obtaining energy for growth and reproduction. Cell biology studies cells in terms of their molecular makeup and the way they cooperate with each other to form very complex organisms, such as human beings.

In order to understand how the healthy human body functions, how it develops, ages and what goes wrong in case of disease, it is essential to know the cells that constitute it. Cell biologists, equipped with increasingly powerful microscopes, proceeded to study the microscopic anatomy of the cell and observed that it is made up of the following structures:

Cell wall: it is a thin covering of non-living materials that covers the cell. In plants this is called cellulose and is characterized by being rigid, in animals this covering is more flexible.

Cell membrane: following the cell wall, this consists of three layers, a middle layer formed by fats (lipids) and two layers of proteins. This structure allows to select which elements can enter or leave the cell.

The cytoplasm: comprises the entire volume of the cell, except for the nucleus. It includes numerous specialized structures and organelles, which will be described later. The concentrated aqueous solution in which the organelles are suspended is called the cytosol; many of the most important cellular maintenance functions, such as respiration, protein synthesis, and photosynthesis, take place in the cytosol. The following organelles are found in the cytoplasm:

The mitochondria: are the centers of respiratory activity, they can break down organic compounds into carbon dioxide and water, which

are exhaled when breathing; when this happens energy is released in the form of ATP. They are particles of 0.2 to 0.3 microns, formed by an outer and an inner membrane with many folds called ridges, which increase the internal surface where ATP production takes place. Mitochondria are abundant in cellular regions where more metabolic activity or energy is required such as in muscle cells or liver cells.

Chloroplasts: the largest particles in cells, they have been found only in plants and some protists. Like mitochondria, they have numerous internal membranes. Photosynthesis takes place in these membranes, since a pigment called chlorophyll is stored here.

Golgi complex: a group of flattened, sac-like membranes located near the nucleus. It is thought to be involved in the excretion and transport of particles in and out of the cell.

Lysosomes: these particles are smaller than mitochondria and contain membrane-enclosed enzymes, which act as catalysts in breaking down large molecules of fats, proteins, and nucleic acids into smaller molecules that can be used as energy sources.

Endoplasmic reticulum: it is a membranous system that connects to the nuclear membrane, so it is possibly involved in internal transport functions of the cell.

Ribosomes: tiny and numerous particles, 0.02 microns in diameter, attached to the endoplasmic reticulum, or free in the cytoplasm. This is where protein synthesis takes place.

Centrioles: rod-shaped structures located near the nucleus of animal cells. They are involved in cell division.

Nucleus: the largest and most important structure in almost all animal and plant cells; it is spherical and measures about 5 pm in diameter. The nucleus carries out a variety of functions:

Reproduction or cellular continuity.

Synthesis of nucleic acids (DNA and RNA).

The nucleus is formed by the following structures:

The nuclear membrane, which surrounds the nucleus and separates its contents from the cytoplasm, has pores through which substances are exchanged between the nucleus and the cytoplasm.

Chromosomes, seen as granules scattered around the nucleus, are difficult to identify separately; but just before the cell divides, they condense and become thick enough to be detectable as independent structures, usually arranged in identical pairs. The nucleolus is a rounded structure, made of RNA, which disappears during cell division. RNA and proteins are synthesized in the nucleolus and migrate into the cytoplasm through the nuclear pores, where they are then modified to become ribosomes.

Processes to which drugs are subjected in the body (pharmacokinetics).

The term pharmacokinetics was introduced in 1953 by the German pediatrician FH Dost, who defined it as "the science of quantitative relationships between the organism and the drug".

Pharmacokinetics studies the processes of absorption, distribution, metabolism and excretion of a drug, determining how often, in what amount, in what dosage form and for how long it must be administered to reach and maintain the required plasma concentrations.

Absorption: is a process of pharmacokinetics that involves the entry of the drug molecule into the body, from its initial site of administration to reach the systemic circulation.

Therefore, it is very important to understand the mechanisms of transport of drug molecules across biological membranes, which constitute barriers to their passage through the cells of different tissues.

There are several factors that condition the absorption as well as circumstances that can alter this process. The route of administration, in turn, imposes certain special characteristics that determine changes in the absorption process.

Mechanisms of transport through membranes.

Unlike the intravenous route, where by definition there is no absorption process, any route of drug administration places the drug molecules in an initial place where they must enter into solution with the liquids of the medium in which they are found. Any movement of the drug molecules within the organism must take place through the biological membranes, which are none other than the plasma membranes of the cells that make up the various tissues.

The plasma membrane of all cells consists of a bilayer of lipids and proteins. Lipids are amphipathic molecules, with their polar or hydrophilic portion oriented towards the outside of the membrane and their apolar or hydrophobic portion, oriented towards the inside of the lipid bilayer. Thus, the passage of a molecule through a cell membrane must face two polar media separated by an apolar one, which thermodynamically represents energy barriers that oppose its crossing.

We will review three types of transport across membranes: passive transport, active transport, and filtration.

Passive transport.

Passive transport is characterized because the driving force of the process to overcome the energy barrier imposed by the lipid bilayer to the drug molecules, is obtained from the formation of an electrochemical gradient (electro by the gradient of molecules that present net electric charge and chemical by the gradient of concentration of the molecules). Thermodynamically, the gradient favors the dissipation of its stored energy in the direction and sense that goes from higher to lower molecular concentration and electric charge of the molecules of the drug. Therefore, it is a process that energetically, does not represent any expense for the cellular economy.

Simple diffusion

Simple diffusion is the simplest form of passive transport.

Most of the drugs, due to their low to medium molecular weight, can cross the lipid bilayer overcoming its energy barrier in favour of the electrochemical gradient. The absence of electric charge of the molecule is a very important factor, since an apolar molecule diffuses more easily than a polar molecule of the same characteristics. However, the degree of polarity of drug molecules is not an absolute value, but depends on the pH of the medium in which the molecule is found.

Most drugs are weak electrolytes, both acids and bases, and in the aqueous medium of the body they are partially ionized. The degree of ionization will depend on the Ka of the molecule and the H+ concentration of the medium: Ka - [H+] = {[A-/ [AH]} /[AH]}

pH = pKa + Log {[A-] / [AH]}

Known Henderson-Hasselbach equation (for a debit acid). The non-ionized form of the molecule (AH, in the example of the debit acid) will diffuse freely through the lipid bilayer depending on its liposolubility. On the contrary, the ionized form (A- in our example), will be thermodynamically difficult to cross the cell membrane due to the presence of electric charge that interacts with the water dipoles. Then we can conclude that if a semi-permeable membrane, such as the cell membrane, separates two compartments (extra and intracellular, for example) in which the drug molecules are dissolved, only the non-ionized form will be able to diffuse across it until the equilibrium state is reached in which the concentrations of the non-ionized form on both sides of the membrane are equalized.

Distribution.

By this process, the drug reaches the body through the bloodstream into the extravascular fluid, either reversibly (distribution) or irreversibly (elimination).

Drugs can be found in the circulation mainly in free form or bound to plasma proteins and haematite, and a balance is reached between one form and the other. It is important to know that only the free form can diffuse from the blood to the sites of action and interact with them to produce a pharmacological effect, reach storage sites or elimination systems.

As a rule, the mechanisms governing distribution are the same as those governing absorption. Let us remember that the most important one is passive diffusion. Therefore, those drugs with high lipophilicity will diffuse more easily through the tissues and will have more facility to cross the different membranes of the organism. Such is the case of psychotropic drugs, general anaesthetics, etc.

Biotransformation Processes

Drugs are eliminated from the body by 2 fundamental mechanisms: hepatic metabolism and renal excretion.

Water-soluble drugs are generally excreted in unmodified form from the kidney, but fat-soluble drugs are not, because when they are filtered by the glomerulus, they are reabsorbed, due to their fat-solubility, by the proximal tubule. Through metabolism, drugs are transformed into more polar, more water-soluble substances. This takes place mainly in the liver by means of two types of chemical reactions: phase I or non-synthetic and phase II or synthetic.

Excretion

Excretion is the process by which a drug or metabolite is eliminated from the body without further modification of its chemical structure.

Drugs or their metabolites are eliminated from the body by 2 fundamental mechanisms: hepatic elimination (the drug is metabolized in the liver and excreted by the biliary tract) and renal excretion (drugs can be removed from the circulation by glomerular filtration or active tubular secretion or passive tubular reabsorption).

Renal excretion

Glomerular filtration. This is the most common route of renal elimination. The free form of the drug is eliminated by filtration, while the protein-bound form remains in the circulation, where some of it dissociates to restore equilibrium.

Conclusions:

Cellular and molecular biology is very important in the knowledge of the processes that drugs undergo in their passage through the body to make their pharmacological effect and be eliminated from the body.

Topics from the Chemistry syllabus:

I: Structure of substances.

II: Water and dispersed systems.

III: Colligative properties.

IV: Thermodynamics.

V: Chemical kinetics.

VI: Chemical equilibrium

VII: Generalities of organic chemistry.

VIII: Hydrocarbons.

IX: Other Organizational Functions.

X: Carbohydrates.

XI: Lipids.

XII: Aminoacids and proteins.

XIII: Nucleic acids.

For each topic, students are shown learning situations where a disease related to the content of the topic is described in which general aspects related to the disease are offered such as: incidence in the population, molecular causes that provoke it, symptoms and signs and general treatment; to make reflections with the students about their opinions about the contradictions implicit in the situation exposed and that can be related to the chemical content. These learning situations constitute simulations that in the area of health, consists of placing a student in a context that mimics some aspect of reality and to establish, in that environment, situations or problems similar to those that must face with healthy or sick individuals, independently, during the different clinical practices.

Clinical simulation is an educational tool that promotes the acquisition of certain technical skills and competencies necessary for health care.

One of the big differences between traditional and simulation-based medical education is that during clinical training on real patients, students must be continuously supervised to avoid making mistakes and correct them immediately, in order to take care of the integrity and safety of the patient; in contrast, within a simulation, mistakes are allowed by the instructor, so that the student learns the consequences of his mistake, rectifies and re-performs the procedure correctly, thus reinforcing his knowledge.

Examples of learning situations:

Theme XI: Lipids

An example of a learning situation that can be used for this content is the following:

Male patient, seven years old, with a height of 1.20 m and weight of 58 kg, is brought to the polyclinic for presenting increased heart rate. Physical examination revealed fine features with double chin, large adiposity in the mammary region and abdomen. A thorax x-ray showed an increase in the size of the cardiac silhouette and the results of complementary tests showed a slightly increased level of triglycerides. The interview showed that she was fatigued with light physical exercise, however, two students who were present in the on-call team as part of their pre-professional practice considered the situation as normal.

What is the explanation of this situation, taking into account the structure-property-function relationship of lipids as biomolecules?

Questions related to the learning situation:

1. Analyzes the patient's lab results, symptoms and signs and identifies associated disease
2. Why are this patient's triglyceride levels elevated?
3. If triglyceride levels are reduced, how is it reflected in the patient's health status?
4. How can these levels in adipose tissue be reduced?

Do you consider that in the situation you have proposed, the structure-property-function relationship is no longer fulfilled?

Triacylglycerides are the lipids responsible for the thickening of adipose tissue. About these say:

a. Family of lipids to which they belong.
b. Chemical composition.
c. Main functions.
d. After having dabbled in the study of lipids, how do you explain the relationship of each of them with obesity?

5. Explain the relationship between the symptoms and signs present in the child with the presence of saturated fatty acids in the triglyceride structure.
6. An excessive intake of triglycerides in the diet can cause serious health complications, however these lipids are very important in the body. Justify the above statement.
7. What are the implications for the child's health if this condition is not

treated in time?

The enunciated learning situation is applicable to any career and must be adapted to the learning needs of the students in each one of them. In the same way, the questions related to this one can be used, as they are in function of the formation of the health professional, in this case, a technologist.

Theme XII: Amino Acids and Proteins

Patient of 10 months of age who comes to the pediatrician for presenting: delayed psychomotor development, delayed dentition, low weight and height for age, aggressive behaviors and poor learning ability.

When the corresponding study was carried out, taking into account the symptoms, phenylketonuria was diagnosed. In an interview with the family members, they reported that the Guthrie test was performed within the established time and the result was negative. Explain the case presented from your professional point of view.

Questions related to the learning situation:

1. Describe the Guthrie test.
2. Identify according to the technique used and established the possible sources of error.
3. Normal blood phenylalanine values.
    a. What type of biomolecule is phenylalanine.
    b. Describe its structure.
    c. Explain the relationship of the biomolecule to disease.
4. Main characteristics and functions of the instruments used in the development of the technique.
5. Other tests that may be performed on the patient to confirm the diagnosis.

Screen galena of images

It shows a similar composition to the previous screens, since it has the Articles and Galenas buttons. When clicking on the Galenas button options, a content box appears in which there is an image viewer, which displays an enlarged image relating to product topics, on the right side of the content box are thumbnails of images, which when clicked are displayed enlarged in the viewer box and its description is displayed in the box below.

Images:

1. Schematic of the molecular structure of HCG.
2. Structure of the plasma membrane.
3. Receptor structure.
4. Membrane surface receptors.
5. Membrane receptors.
6. Nature of receptors: transporters and ion channels.
7. Nature of the receptors: enzymes.
8. Nature of the receptors: physiological.
9. Mechanism of enzyme inhibition.
10. Reve rsible inhibition.
11. Irreversible inhibition.
12. Transporter proteins.

13. Sodium-potassium pump.
14. Types of channels.
15. Transport through a channel.
16. Receiver coupled to ion channel.
17. G protein-coupled receptor.
18. Receptor with enzymatic activity.
19. Receptors that regulate gene transcription.
20. Receptors associated with ion channels (ionotropic).
21. Ligand-regulated ion channel.
22. G protein-coupled receptors.
23. Receptors with enzymatic activity.
24. Nucleus-acting membrane receptors
25. Receptors that regulate gene transcription.
26. Receptor regulation: desensitization or downregulation.
27. Inhibition of the synthesis of new receptors.
28. Modifications in the efficacy of the receptor_effector system.
29. Benzodiazepines.
30. Importance of receptor subtypes.
31. Non-receptor-mediated drug actions.
32. Mechanism of spurious incorporation.
33. Diseases resulting from receptor dysfunction.

A. Myasthenia gravis.
B. Pseudohypoparathyroidism.
C. Neoplastic diseases.

D. Insulin-resistant diabetes mellitus.

E. Retinitis pigmentosa.

F. Testicular feminization syndrome.

34. Structure of saturated and unsaturated fatty acids.
35. Action of bile acids on fats.
36. Structure of triacylglycerols.
37. Structure of glycerol phosphatides.
38. Structure of sphingolipids.
39. Structure of vitamin E.
40. Foods containing vitamin E.
41. Importance of vitamin E.
42. Structure of vitamin K.
43. Sources of vitamin K.

Galena video screen

It shows a similar composition to the previous screens, since it has the Articles and Galenas buttons. When clicking on the Galenas button options, a content box appears in which there is a video viewer, which shows a video about the product topics; on the right side of the content box there are thumbnails of videos, which when clicked are displayed in the viewer box and its description appears in the box below.

Below the video playback box are the control buttons. At the bottom right is the back button, which when clicked returns to the Galenas screen.

Videos:

1. Introduction to Educational Chemistry.
2. Basic Chemistry.
3. What is Chemistry. Everything is chemistry.
4. General Chemistry Tutorial.

The application of the software in the process of teaching and learning of chemistry linked to biomedical specialties for students and health professionals, allows a new methodological vision of the process from its conception from the organization of cognitive activity that integrates coherently the content of the science of chemistry with its practical applications.

RESULTS

The following results were obtained from the initial diagnosis:

From a total of 45 students surveyed, 100% agreed that the bibliography for learning is insufficient and there are not enough computer media to train students in the content of this science.

In this initial diagnosis only 13 students were able to cite examples of the link between their speciality and the chemical contents for a 28.8% pass rate, which shows the need for teaching support materials such as computer resources. This has a proportional influence on the results of the two aspects measured below.

Of the 10 teachers interviewed, 9 said that they do not use new methods for 90% because they do not have a specialized bibliography and also insufficient computer resources on which they can rely to facilitate the learning of these contents.

The methodological guidelines are insufficient to achieve this linkage, although they guide what needs to be done.

RESULTS OF THE APPLICATION OF THE EXPERT JUDGEMENT

METHOD

When 20 experts were consulted, an estimation error of 1% (a= 0.01) was introduced, so it can be affirmed that the decisions taken, based on the calculations made, are highly reliable and valid.

In this case the methodology of preference was selected, allowing to achieve a high level of objectivity and accuracy, combined with the speed and ease of statistical tabulation of the results, without sacrificing the validity of the judgment derived from its application.

This methodology establishes a balance between the level of complexity of application, processing of the obtained data and validity of the results, which allows to reach an integral and wider image of the possible evolution of the analyzed scientific result.

A survey with 6 item questions and one development question was applied: The first 6 answers had to be associated with the following qualitative criterion:

6 - Totally agree.

5 - Strongly agree.

4 - All right.

3 - Neither agree nor disagree.

2 - Disagree.

1 - Strongly disagree.

The following aspects were submitted to expert judgement:

- The theoretical underpinning of the technological alternative.
- Relevance and possibility of the technological alternative for linking chemistry with biomedical sciences.
- Actuality and novelty of the technological alternative.
- Linkage of the content treated and organized in the proposed technological alternative with respect to the professional profile of the graduates of the careers of Health Technology.
- Impact of the proposed technological alternative in the learning of

biomedical sciences in students of Health Technology careers.

The results of the ordering carried out by each expert to the different aspects of the guide show that the elements submitted to their criteria were evaluated, by all, between 4 and 6 (totally agree, strongly agree and agree), which demonstrated a high degree of acceptance by the specialists consulted on the theoretical and methodological conception of the proposed alternative and, therefore, the feasibility of applying it in educational practice with presumably satisfactory results.

The remarks made by the experts in the answer to the development question were taken into account in the improvement of the technological alternative, and in fact, the changes referred to them were taken into consideration in the proposal that was applied in the pedagogical practice.

The educational software developed was approved by expert criteria and put into practice for the students of Health Technology, which allowed an adequate linking of Chemistry with the Basic Biomedical Sciences and favored the learning of the chemical contents in their practical applications in the field of health.

RESULTS OF THE APPLICATION OF THE SOFTWARE IN PRACTICE EDUCATIONAL

The modality of experiment applied was the quasi-experiment, for it the control group was conformed by 23 students, 11 of Clinical Bioanalysis and 14 of Hygiene, the experimental group was conformed by 22 students of Imaging.

The following dimensions were designated for the test that was applied at the end of the experiment:

- Cite examples of chemical substances used in the profession.
- Explain the structure-properties-functions relationship of substances.
- To exemplify the application of the chemical content, taking into account

the structure-properties-functions relationship in their professional practice.

It is possible to appreciate the superior results achieved by the experimental group in the three dimensions, also, if the results are compared, it is evident that the most difficult is to exemplify the application of the chemical content, taking into account the relationship structure properties functions in their professional practice, both for the experimental (95.9%) and for the control (43.4%). (Table 1).

Table 1. Results of the quasi-experiment

Dimensiones	Grupo control	%	Grupo experime	%
1	15	65.2 %	20	100%
2	10	52.17 %	22	100%
3	12	43.4%	22	95.9%

The evaluation of the results of the quasi-experiment is carried out in two ways, the first, referred to the comparative elements between the learning results of the students in the experiment of verification with respect to the results derived from the initial diagnosis. In this case it was appreciated, both in the control group and in the experimental group, advances in the learning level; in addition, it was achieved that the students arrived to conceptual definitions and development of specific abilities, when explaining the relation structure-properties-functions and the link with the professional practice, in the space of incidence of the health technologists, elements of knowledge depressed at the moment of beginning of application of the alternative.

In a second sense, a growth was achieved in the experimental group as the sum of students located in high and medium level reaches 93.75% of the total by 83.79% in the control group, results that show a qualitative leap in the learning of students in the experimental group, in which it was achieved that most were located in a high level of learning to exemplify the application of chemical content, taking into account the relationship structure-properties-functions in their professional practice.

The results of the application of the quasi-experiment allowed to verify the positive aspects derived from the application of the proposal in the process of teaching-learning of the chemical contents.

DISCUSSION

The implementation in the educational practice of the technological alternative allowed to raise the level of learning of the chemical contents in their practical applications in the field of health in the students of the careers of Health Technology.

The increase in the level of learning after the investigation is based on the fact that the assimilation of the contents by the students will be increased by 50 %.

The social meaning of the same is favoured if it is understood, hence the importance of its link with the exit of the professional.
Jaramillo, Castaneda and Pimienta, quoted by Cruz Perez and others [(34)] state that: "They allow the implication of the student in their tasks and develop their initiative, since they are constantly forced to make decisions, to filter information, to choose and select, improve the attitudinal performance in the subject, allow establishing dialogue and discussion that contribute to improve learning, participate more: question, analyze, argue and propose, reflect in their contributions what they are understanding, allowing their evaluation".
Several researches have been developed on the use of ICT in the process of teaching and learning of chemistry, among which we can mention:
Using ICT to teach Chemistry to deaf students. Primera experiencia de integration de alumnos sordos en Ensenanza Media de Berrutti, en el ano 2008[(35)]; Recursos tic para el aprendizaje de la Quimica y la Fisica en el ciclo basico universitario de Vera y otros en el ano 2018 [(36)]; La virtualization del contenido nomenclatura quimica en la educación superior pedagogica de Mesa, Blanco y Addine en el ano 2018 [(37).]
None of these studies addresses the process of teaching and learning of chemical contents in their practical applications in the field of health. Chemistry allows the study of life at the molecular level, it is a field of enormous scientific interest and of vital importance for the professionals linked to the medical sciences, being necessary that the educators introduce in the process of teaching and learning new methods and means that allow its better understanding.
In the present investigation, a limitation was the unavailability of sufficient computers for the use of the product by the students of the experimental group, in the established period. It was necessary to carry out the experiment in hours outside teaching hours.
Methodological activities were developed with the group of teachers to guide them in terms of types of classes in which the product could be used

and how to use it attending to the phases of orientation, execution and control for the organization of the cognitive activity of the students.

The structure of the technological alternative guarantees the interactivity of the student with the computer, to the extent that it operates with specific buttons for each computer action, it is aided by sound and locutions that activate the hearing as an important sensory organ in learning. The soundtrack is designed to provide a pleasant environment for the learning of chemical contents that are abstract and complex for the comprehension by the students.

CONCLUSIONS

It can be concluded that the application of the software in the process of teaching and learning of chemistry, linked to biomedical specialties for students and health professionals, allowed a new methodological vision of the process from its conception from the organization of cognitive activity that coherently integrates the content of the science of chemistry with biomedical specialties.

The implementation in the educational practice of the technological alternative, after demonstrating its feasibility, allowed to raise the level of learning of the chemical contents in their practical applications in the field of health in the students of the careers of Health Technology.

BIBLIOGRAPHIC REFERENCES

1. Alvarez Valcarcel JL, Sanchez Camacho Z. Quimica biomoleculas [Internet].Havana: Ecimed; 2017[cited 2/4/2019]. Available from: http://www.bvscuba.sld.cu/libro/quimica-biomoleculas/
2. Alcocer Aparicio PM, Rodriguez Morales A, Arango GonzalezJL. Formation docente para promover valores morales en la Universidad de Guayaquil.Universidad y Sociedad [Internet]. 2016[cited 7/9/2019];8(2):[approx. 7 p.]. Available from:http://scielo.sld.cu/scielo.php?script=sciarttext&pid=S2218-36202016000200024&lng=en&nrm=iso>.
3. Isabel S. Benefits of technology in education[Internet]. Santiago, Chile: U-planner.com; 2016[cited 26/12/2018]. Available from: https://www.u-planner.com/es/blog/beneficios-de-la-tecnolog%C3%ADa- en-educaci%C3%B3n.
4. Rodriguez BeltranNM. Formative dynamics in telemedicine for the careers of Medical Sciences [Internet]. Santiago de Cuba: University of Oriente. Centro de Estudios de Educaci6n SuperiorManuel F. Gran; 2014[cited 13/2/2019]. Available from: http://tesis.sld.cu/index.php?P=FullRecord&ID=193
5. Rodino Hoyos CA. Utilization of ICT as a didactic strategy to facilitate the teaching-learning process of chemistry in the tenth grade of the superior normal school of Monterrey Casanare [Internet]. Colombia: Universidad Nacional Abierta y a Distancia (UNAD); 2014[cited 26/12/2018].Available from: https://stadium.unad.edu.co/preview/UNAD.php?url=/bitstream/10596/2688/1/7382890.pdf.
6. C. the Caribbean. [Internet] Republics Domicana: C. the Caribbean. 2015

[cited 13/2/2019] Mendez AlonsoLM. Model in the Teaching of Chemistry supported in ICT: Resources and experiences of its correct application; [approx. 18 p.]. Available from: https://www.elcaribe.com.do/2015/08/27/modelo-educacion-ensenanza-quimica-soportado-las-tic- recu rsos-experiencias-correcta-aplicacion/.

7. Perez MartinotM. Current use of information and communication technologies in medical education.RevMedHered[Internet]. 2017[cited 13/2/2019];28:[approx. 4 p]. Available from: http://web.b.ebscohost.com/ehost/pdfviewer/pdfviewer?vid=0&sid=d5a49e90-64b4-4bd0-a231-dbc747a209a9%40sessionmgr103.

8. Garces Suarez E, Garces Suarez E, Alcivar Fajardo O. The technologies of lainformation in the change of higher education in the 21st century: reflections for practice. University and Society. [Internet]. 2016 [cited 13/2/2019];8(4): [approx. 7 p.]. Available from: http://scielo.sld.cu/scielo.php?script=sci arttext&pid=S2218-36202016000400023.

9. ICTs in the teaching of chemistry. 2019[cited 26 Dec 2018] In: Bilogiauniversitaria.blogspot.com [Internet]. Cuba: Biochemistry and Education. [approx. 2 screens]. Available from: https://bilogiauniversitaria.blogspot.com/2019/02/las-tics-en-la-ensenanza- de-la-quimica.html.

10. Huarancca Rojas E. Application of the dialectical method in the development of research skills [Internet]. Peru: 3ciencias. Editorial Area de Innovation y Desarrollo, S.L; 2020 [cited 26 Dec 2020]. Available from: https://doi.org/10.17993/DideInnEdu.2020.48

11. Soviet dictionary of philosophy [Internet]Spain: Filosofia en espanol project[cited 26 Dec 2020]; 2017. Gnoseologia. Available from: http://www.filosofia.org/enc/ros/gnos.htm

12. Macias Alvia A, Hurtado Astudillo JR, Cedeno Holguin DM, Vite Solorzano FA, Scott Alava MM, Vallejo Valdivieso PA. Introduction to the study of Biochemistry [Internet]. Peru: 3ciencias. Editorial Area de Innovation y Desarrollo, S.L; 2018 [cited 26 Dec 2020]. Available from: http://dx.doi.org/10.17993/CcyLl.2018.28

13. Yera Quintana AI, Perez Hernandez I, Rodriguez Garcia LR. Process of teaching and learning of chemistry in link with the locality. Sustentos de partida. Education y Sociedad[Internet] 2020[cited 20 Nov 2020]; 18(3): 1-15. Available from: http://revistas.unica.cu/index.php/edusoc/article/view/1651/html

14. Vergara Vera I. Dynamics of the teaching-learning process of chemistry in health technology. Education Medica Superior [Internet]. 2014 [cited13/2/2019];28(2):[approx. 11 p.]. Available from: http://ems.sld.cu/index.php/ems/article/view/258

15. Silva Quiroz J, Maturana Castillo D. A proposed model for introducing active methodologies in higher education. Innov. educ [Internet]. 2017 [cited 23 Jan 2020]; 17(73): 117-131. Available from: http://www.scielo.org.mx/scielo.php?script=sciarttext&pid=S1665-26732017000100117&lng=en&nrm=iso>. ISSN 1665-2673.

16. Gonzalez La Nuez O, Suarez Suri G. The means of teaching in the special didactics of the discipline Human Anatomy.

Rev.Med.Electron. [Internet]. 2018 Aug [cited 25 Jan 2020] ; 40(4): 1126-1138. Available from: http://scielo.sld.cu/scielo.php?script=sci arttext&pid=S1684-18242018000400018&lng=es.

17. Psicopedagogia.com [Internet]. Madrid: Psicologia de la education para padres y profesionales; 2006 [cited 20 Jan 2020] Lombillo Rivero I. The use of teaching media and their influence on students' creativity. Available at: https://www.psicopedagogia.com/medios- ensenanza

18. Peraza Zamora C, Gil Lopez Y, Pardo Garcia Y, Soler Cruz LO. Characterization of the teaching means in the teaching-learning process in Physical Education. PODIUM [Internet]. 2017[cited 20 Jan 2020]; 12(1):4-11. Available from: http://podium.upr.edu.cu/index.php/podium/article/view/681/html

19. Capanegra HA, Cabrera G, Aguilar ML, Jorda MS. The use of information and communication technologies (ICTs) in the university environment. Documentos y Aportes en Administration Publica y Gestion Estatal [Internet]. 2016[citado29/4/2019]; 16(26):159-90. Available from: http://search.ebscohost.com/login.aspx?direct=true&db=a9h&AN=118657851&lang=en&site=ehost-live.

20. Fernandez Silva IL. Pedagogical model for the use of the computer in the educational attention to schoolchildren with mental retardation [Internet] Havana: University of Pedagogical Sciences "Enrique Jose Varona" Faculty of Child Education Department of Special Education; 2013[cited 20 Jan 2020]. Available from: https://www.google.com/url?sa=t&rct=j&q=&esrc=s&source=web&cd=&cad=rja&uact=8&ved=2ahUKEwjdi6XLnbuAhWKjFkKHfaUBkQQFjAFegQICxAC&url=http%3A%2F%2Feduniv.reduniv.edu.cu%2Ffetch.php%3Fdata%
3D508%26type%3Dpdf%26id%3D508%26db%3D1&usg=AOvVaw0miMNEiXZsr6aF1ubF9ed

21. Castro Martmez Jesus Alejandro, Gonzalez Lorenzo Liena, Gonzalez Lorenzo Leiny. Considerations on community social psychology with a Marxist vision in primary health care. Medicentro Electronica [Internet].

2020 [cited 20 Jan 2020] ; 24(1): 129-148. Available from: http://www.medicentro.sld.cu/index.php/medicentro/article/view/2707/2510

22. Arce Hernandez M, Tellena Prieto M del C, Barrios Poo W, Morejon Alfonso M, Arce Puente D. The impact of the multimedia "Cochlear Implant" in the educational teaching process of medical sciences. Rev. Medical Sciences [Internet]. 2015 [cited 20 Jan 2020];19(2): [approx. 7 p.]. Available from: http://www.revcmpinar.sld.cu/index.php/publicaciones/rt/printerFriendly/2028/html_65.

23. Unir. La universidad en internet [Internet]. Spain: International University of La Rioja, 2020 The zone of proximal development and its application in the classroom [updated 28 Aug 2020; cited 20 Jan 2020]. Available from:_https://www.unir.net/educacion/revista/zona-desarrollo- proximo/

24. Cardenas Ramos RA, Lopez Lanayaco DA. Interactivity in social networks and the processes of teaching-learning of the students of the seventh cycle of the educational institution Tupac Amaru of Tapuc Yanahuanca 2017 [Internet]. Cerro de Pasco: Universidad Nacional Daniel A. Carrion;2018 [cited 9 Jul 2020]. Available from: http://repositorio.undac.edu.pe/handle/undac/360

25. Monsalve Castro NY, Monsalve Castro C. The inclusion of the computer in the classroom by teachers of fifth grade of elementary school as a tool to promote meaningful learning in students. Rev. esc.adm.neg [Internet] 2015[cited 20 Jan 2020]; 79: 50-63.
Available at: https://www.redalyc.org/pdf/206/20643042004.pdf

26. Vitor Maldonado RM. Educational software in the development of learning at the initial education level. [Internet] Lima: National University of Education, Enrique Guzman y Valle; 2017[cited 20 Jan 2020]. Available from:
https://repositorio.une.edu.pe/bitstream/handle/UNE/4426/Los%20softwar

e%20educativos.pdf?sequence=1&isAllowed=y.

27. Rodriguez Lamas R. Introduction to Educational Informatics. Havana: Instituto Superior Politecnico Jose A. Echevarria; 2007

28. Castaneda Urdaneta FA, Chiang Borges JR, Lobaina Rosales OM, Panadero Vega RM, Vega Sanchez M del C, Rodriguez Hung S. CiruSoft, multimedia para el estudio de la asignatura Cirugia.Holguin: Edumed Holguin; 2019 [cited 20 Jan 2020]. Available from: http://edumedholguin2019.sld.cu/index.php/2019/2019/paper/viewFile/231/ 151

29. Rojas C, Garcia L, Alvarez A. Methodology of teaching chemistry. Havana: Editorial Pueblo y Education; 2002.

30. Silvestre Oramas M, Zilberstein Toruncha J. Hacia una didactica desarrolladora. Havana: Editorial Pueblo y Educacion; 2002.

31. Ausubel D; Novak J; Hanesian H. Psicologia Educativa: un punto de vista cognoccitivo. Mexico: Editorial Trillas; 1986.

32. Rodriguez Lamas R. Introduction to Educational Informatics. Havana: ISPJAE; 2000.

33. FigueredoTrimino N, Garcia Leyva L, Perez Matos RW. The teaching-learning of university General Chemistry with the use of professionalized teaching tasks. Teacher and Society [Internet] 2018 [cited 20 Jan 2020] 15(4): 603-617. Available from: https://www.google.com/url?sa=t&rct=j&q=&esrc=s&source=web&cd=&cad=rja&uact=8&ved=2ahUKEwiny Sm88PuAhWBxFkKHR7lAjAQFjACegQIBBAC&url=https%3A%2F%2Fmaestroysociety.uo.edu.cu%2Findex.php%2FMyS%2Farticle%2Fview%2F4391%2F3769&usg=AOvVaw3RR4pGyh C4bY0XaloRjyX

34. Cruz Perez MA, Pozo Vinueza MA, Andino Jaramillo AF, Arias Parra AD.

Information and communication technologies (ICT) as an interdisciplinary research form with an intercultural approach for the students' formation process. Eticanet [Internet] 2018 [cited 26 Dec 2020]; II(18): 196-215. Available from: https://www.google.com/url?sa=t&rct=j&q=&esrc=s&source=web&cd=&cad=rja&uact=8&ved=2ahUKEwjwhJfHubzuAhWNuVkKHQd-CkkQFjAAegQIARAC&url=https%3A%2F%2Fdialnet.unirioja.es%2Fdescarga%2Farticulo%2F6840740.pdf&usg=AOvVaw0MBxH2XnR839jJ XQqG 2w6

35. Berrutti S. Relying on ICT to teach chemistry to deaf students. [Internet]Uruguay: Liceo N° 32 de Montevideo; 2008 [cited 29/4/2019]. Available from: http://www.niee.ufrgs.br/eventos/SICA/2008/pdf/C109%20Quimica.pdf

36. Vera MI, Lucero I, Stoppello MG, Petris RH, Gimenez LI. Resources tic for the learning of Chemistry and Physics in the basic university cycle. In: XX Workshop of Researchers in Computer Science [Internet]; 2018 Argentina [cited29/4/2019]. Available from: http://sedici.unlp.edu.ar/bitstream/handle/10915/68682/Documento compl eto.pdf-PDFA.pdf?sequence=1&isAllowed=y.

37. Mesa Brinas GH, Blanco Gomez MR, Addine Fernandez R. The virtualization of chemistry nomenclature content in higher education pedagogy. RITI [Internet]. 2018 [cited 29/4/2019];6(12):[approx.10 p.]. Available from: http://www.riti.es/ojs2018/inicio/index.php/riti/article/view/116/html

Printed by Books on Demand GmbH, Norderstedt / Germany